A BEGINNER'S GUIDE TO
Astronomy

Alistair Glasse

CAXTON EDITIONS

About the Author
Alistair Glasse holds a Ph.D in Infrared Astronomy from
the University of London, and is currently a Project Scientist at
the Royal Observatory, Edinburgh.

First published in the UK in 1999 by
Caxton Editions
16 Connaught Street
Marble Arch
London W2 2AF

Copyright © Mars Publishing, 1999

All rights reserved. No part of this publication may be reproduced, stored
in a retrieval system or transmitted in any form or by any means,
electronic, mechanical, photocopying, recording or otherwise,
without the prior permission of the copyright holder.

ISBN 1-84067-088-6

This is a MARS book

Edited, designed and produced by Haldane Mason, London

Editorial Director: Sydney Francis
Art Director: Ron Samuel
Project Editor: Naomi Waters
Picture Researcher: Naomi Waters

Printed in China

Picture Acknowledgements
All pictures supplied by Science Photo Library: 4–5, 20–21 (Jerry Schad);
6 (George Post); 7, 18, 37 (Frank Zullo); 10, 38–39 (Chris Butler); 11, 13B,
26–27, 56, SPL; 12, 16, 17, 29T, 29B, 48, 49, 50, 51, 52, 53, 57, 58, 61, 62, 63
(NASA); 13T (David Hardy); 14, 22, 24–25 (Dr Fred Espenak); 15, 32 (Ronald
Royer); 19 (John Sanford); 23 (Pekka Parviainen); 33 (Royal Observatory,
Edinburgh); 34 (Dr Ian Gatley/National Optical Astronomy Observatories);
35, 59 (Space Telescope Science Institute/NASA); 36L (MSSSO, ANU);
36R (Harrington et al); 40 (Sally Bensusen); 45 (David Ducros);
46T, 46–47 (US Geological Survey); 60 (James King-Holmes).

Contents

Introduction	4
1: Heavenly Bodies	8
The Sun	10
The Moon	16
The Stars	20
The Ecliptic	26
Venus	29
2: The Origins of the Solar System	30
Nebulae	32
Comets, Meteors and Asteroids	36
The Discovery of New Planets	38
3: Exploring the Solar System	42
Exploring in the 21st Century	44
Mars	46
Jupiter	48
Saturn	52
4: Beyond the Solar System	54
Exploring the Universe	56
The Search for Extra-terrestrial Life	60
Index	64

Astronomy

Introduction

Astronomy is a science whose beauty can be appreciated by anyone who can go out on a dark night away from street lights, and look up at the pale ribbon of the Milky Way stretching across the starry sky. It seeks to answer questions which are simple and yet profound. Where did the world come from and how will it end? Does the universe stretch away into the distance forever, and if not, what lies beyond? These are very difficult questions

Introduction

to answer, yet physicists and astronomers have built up a rational picture of the universe we live in, where our own planet Earth has orbited the Sun for the past four billion years since it was formed from a disk of primeval dust and gas. The Sun in turn is an ordinary star like the thousands you can see as distant bright points in the night sky, and it is only one of the many billions of stars which are clustered together in our disk-shaped galaxy. It is the diffuse light of the stars in the galaxy that you see encircling the sky when you look at the Milky Way.

The fundamental questions of existence are still shadowed in mystery, but the desire to search for answers is one of humanity's greatest qualities, and astronomy is continually making discoveries which can fill us with wonder and remind us that our worldly concerns look pretty insignificant when viewed against the vast background of space.

The Milky Way, our own galaxy, seen over a line of trees. The large bright patch is the star cloud M24.

Basic Observing Equipment

My first piece of advice for a young astronomer is this. Don't rush out and buy a telescope. Unless you live in the countryside with a garden which is not flooded by street lighting, the effort of lugging a heavy and delicate telescope to a decent site on a clear night will soon dull your enthusiasm. A good pair of binoculars is ideal for observing the Moon or comets, thanks to their large field of view and portability. For meteor showers, where grains of dust, burning up as they enter the atmosphere, make bright streaks across the sky, your eyes will give you the best view of all.

If you are determined to acquire your own telescope, avoid refractors (ones that use lenses as the primary light collector); they are toys. You should go for a reflecting telescope with the largest aperture you can afford. The aperture is the diameter of the mirror which collects the light and focuses it down onto your eye or the photographic film of a

Amateur astronomers shield their eyes from the glare of the Sun.

camera. The aperture of your unaided eye is equal to the diameter of the dark iris that lets in the light, just half a centimetre or so. A telescope with a 10-cm diameter primary mirror will therefore collect about four hundred times as much light as your eye, allowing you to see four hundred times as many stars. In theory, it will allow you to see about twenty times as much detail as the unaided eye, but in reality you will find that what you see depends a good deal on the observing conditions.

The most important rule for astronomers of any age is never

to go observing at night on your own. Apart from the obvious dangers, observing is much more fun when you are out with a friend or family member, and just on the off-chance that you do discover a nova it helps to have a witness! There are plenty of amateur astronomy clubs and societies around the world keen to recruit new members, and apart from meeting up for observing sessions, they will often invite professional astronomers to talk about their latest research; they may even have access to an observatory with telescopes.

In cooler climates the amateur astronomer's list of essential equipment starts with a warm coat, gloves and a hat. The hat can be as silly as you like because no one will see it in the dark, but it must be warm; ear flaps are recommended. Next on the list are a simple star chart and a torch to read it by, and take a notebook and pencil so that you can make a quick sketch of the stars around the location of anything interesting that you see. You can then identify it properly later. A compass can be useful the first time you go out, to help find which way is north, although you will soon get experienced at finding the north star, Polaris, which we will discuss later.

The Sun analemma, the figure of eight shape made by plotting the position of the Sun at the same time of day during the year.

Heavenly Bodies

- **The Sun**
- **The Moon**
- **The Stars**
- **The Ecliptic**
- **Venus**

CHAPTER 1

Heavenly Bodies

The Sun

In the ancient world the ability to tell the time was vital for farmers, who needed to know when to plant, when to harvest and when the winter would end. The regular motion of the Sun and Moon across the sky provided people with a perfect clock and calendar, and so it is no suprise that nearly all ancient civilizations were expert at predicting key astronomical events – many even built pointers to them in the form of monuments.

Stonehenge

Stonehenge has standing stones which are placed precisely on a line which points to the spot on the horizon where the sun will rise on Midsummer Day. Astronomers try to understand the universe by investigating its physical properties, but the people who built Stonehenge were less interested in finding out why the

Stonehenge on Salisbury Plain, Wiltshire, England. Construction of this prehistoric temple began over 5,000 years ago.

The Sun

Polish mathemetician, Nicolaus Copernicus (1473–1543), who proposed a new picture of the Universe.

Sun rose where and when it did, than they were in being certain that it would do so predictably.

Copernicus

Five hundred years ago, it was the great Polish mathematician Nicolaus Copernicus who first sketched the picture of our astronomical neighbours that explains why Stonehenge works. He proposed that the apparent motion of the Sun and the stars was caused by the Earth moving through space, spinning on its own axis once every day, and then travelling around the Sun in a great orbit once every year. This idea was revolutionary in medieval Europe. It suggested that the Earth wasn't at the centre of the Universe, and posed the question of what it is that stops everything – people, houses and trees – from flying off the Earth into oblivion.

Two hundred years later Isaac Newton showed that this mysterious force was gravity. The force of gravity attracts any two objects which have mass towards each other, and it gets stronger the more massive the objects are and the closer they are together.

Time-lapse photograph of the positions of a setting sun.

Heavenly Bodies

Gravity holds you to the surface of the Earth. The Sun is massive, and the Moon is, in astronomical terms, quite close, and so it is these two bodies which have the strongest pull on our Earth, to the extent that our ocean's tides follow the positions of the Sun and Moon in the sky.

X-ray image of the Sun's corona, rising up to 75,000 km above the Sun's surface.

Solar Energy

It has been one of the great successes of astronomy that we now have a good answer to this question of what makes the Sun shine. Only nuclear fusion could generate the amount of energy that has powered the Sun for the past five billion years, and it will keep it shining for another five billion.

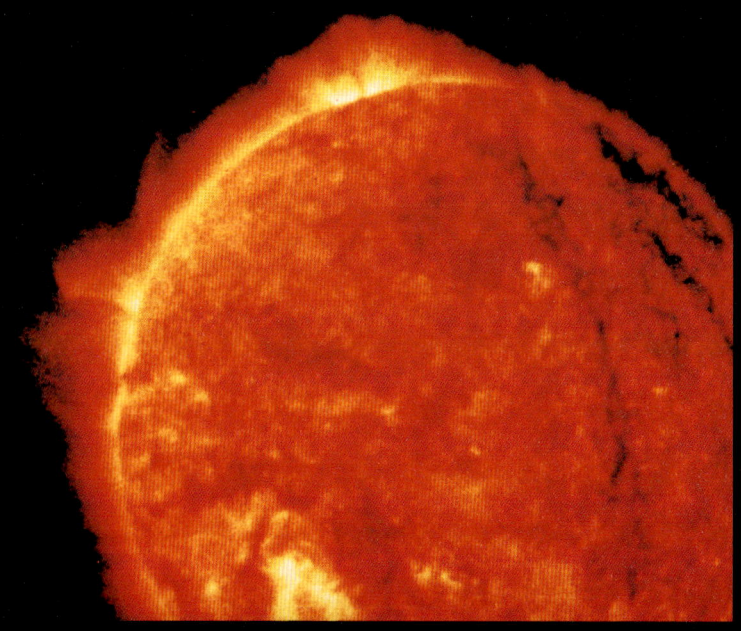

The Sun

At the Sun's core, the nuclei of hydrogen atoms are crushed (fused) together at a temperature of 10,000,000°C to form nuclei of helium. For every 1,000 grams of hydrogen that are burnt in this way, only 993 grams of helium are produced. The missing 7 grams are released as pure energy. If we could harness just this 7 grammes of energy on Earth, it would be enough to hurl a

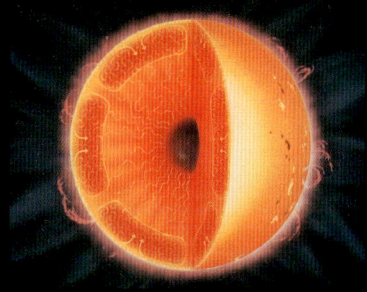

Fusion reactions in the Sun's core create energy which radiates outwards.

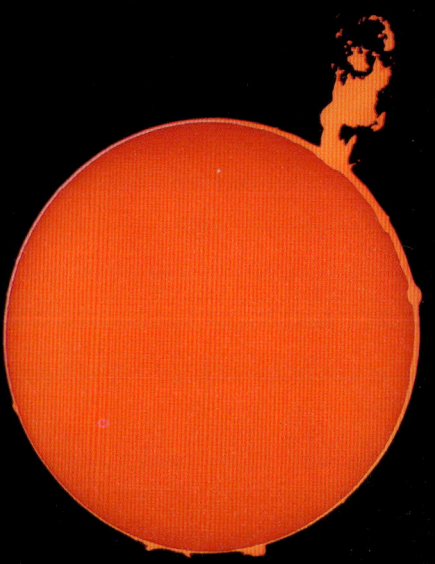

False-colour photograph of a massive solar flare. This photographic technique accentuates the corona and the flare for greater visibility.

Heavenly Bodies

satellite the size of St Paul's Cathedral into orbit. In the Sun's core, four million tonnes of mass are being turned into energy every second.

The corona is the outer layer of the Sun's atmosphere, and is only visible from Earth during a total eclipse. The flares that can be seen streaming away from the Sun are plumes of gas which have been heated to a temperature of tens of millions of degrees in the corona, which then boil away into space. Travelling at hundreds of kilometres per second, the resulting electrically charged solar wind takes just a few days to reach the orbit of the Earth, where it causes electrical storms in our upper atmosphere, disrupting radio communications and lighting up the sky as it gets funnelled down towards the North Pole by the Earth's magnetic field. In spite of its high energy, the solar wind is really only a puny exhaust from the furnace that powers the Sun.

Solar Eclipses

One of the rarest and most spectacular sights in astronomy is a total solar eclipse, which comes about when the Moon, whose silhouette just happens to be the right size, covers the blazing disk

Time-lapse image of a total solar eclipse. The shielding effect of the Moon allows observations of the Sun's corona.

The Sun

of the Sun. This causes darkness during the daytime, and when this occurred in ancient and more superstitious times it was taken as a sign of impending doom and disaster, or a god's displeasure with man's conduct and devotion.

Total solar eclipse. Several prominences may be seen as as red patches in the bright corona. The corona is normally invisible due to the brightness of the Sun's surface.

Heavenly Bodies

The first man on the Moon, Neil Armstrong, seen 'suiting up' for the launch of Apollo 11 on 16 July 1969. The Lunar Module landed in the Sea of Tranquility on 20 July 1969. Stepping out of the Module Armstrong uttered the now famous words, 'One small step for man, one giant leap for mankind.'

The Moon

The Moon is, in astronomical terms, very close (just 380,000 km away), and so despite it's small size compared to the Sun, it still exerts a very strong gravitational pull on our Earth. Our oceans' tides are caused by the Moon, with water accumulating in the two regions of the Earth that are nearest to or furthest from the Moon. Just as the Moon causes tides on the Earth, so the Earth causes tides on the Moon. It is the braking effect of the Earth's gravitational pull on the molten lunar interior that has gradually slowed the spin of the Moon on its axis, so that it is locked into a state where it always presents the same face to the Earth.

The closeness of the Moon has been an inspiration to astronomers for centuries. The mountains and plains that cover its surface and are easily seen from Earth, and the passage of its sunlit phases from New to Full Moon through the month, were alluring to people seeking to understand the heavens. In July 1969 this fascination spurred mankind's greatest

Apollo 11 view the cratered terrain which is typical of the far side of the Moon.

Heavenly Bodies

technical achievement, when the astronauts Neil Armstrong and Buzz Aldrin travelled safely to the Moon on the Apollo 11 mission and walked on its surface.

When you look at the Moon with your unaided eyes, the dark patches show where ancient seas of molten lava froze solid to form wide plains. The light patches are

Time-lapse photograph of the planet Jupiter emerging from behind the Earth's moon. Jupiter is the bright spot on the dark side of the uppermost of the four crescent moons (at upper centre). The phenomenon of three astronomical bodies (in this case the Earth, Moon and Jupiter) being placed along a straight line is known as an occultation or conjunction.

The Moon

View of the full moon, showing its dark 'seas' and two bright rayed craters.

even more ancient mountain ranges rising above the plains. With a pair of binoculars you can see the Moon's circular, rimmed craters, which represent a history of billions of years of meteor impacts on a surface that has been unprotected by its lack of an atmosphere.

Heavenly Bodies

The Stars

he night sky can be very confusing when you first try searching for a particular star or planet, but with a simple chart of the stars (or the inflatable star globe that comes with this book)

The Stars

and a little thought, you will soon be able to find your way around.

Constellations

Stars are very far away and they move across the sky very slowly, so their positions relative to each other take a long time to change. This has allowed people to invent 'join the dots' pictures called constellations out of the random patterns made by bright stars, and once you can recognize just one or two constellations, using them to find the others is easy. If you live in the northern hemisphere and go out on a clear evening in the first few months of the year, you will see the unmistakable, belted-torso shaped constellation of Orion rising above the Eastern horizon. In the summer, when the Earth has moved around to the other side of the Sun, Orion is up during the daytime, and you can't see it. The constellation Ursa Major (which looks more like a great saucepan), is now the one to start off with, or the 'W' shape of the constellation Cassiopea which is nearly always in view, high up in the sky.

The Plough (or Big Dipper) is one of the most prominent constellations, and forms part of a larger group called Ursa Major, or the Great Bear.

Heavenly Bodies

Star trails around Polaris.

One of the most useful stars to be able to find in the north is the Pole or North Star, Polaris. Polaris is very close to the north celestial pole, the point in the northern sky about which all of the stars appear to rotate as the Earth turns. The star trails in the picture above show this clearly. The picture was taken at the Canada-France-Hawaii Telescope at the Mauna Kea Observatory in Hawaii. The camera shutter has been left open for several hours and so the fixed stars have left circular trails across the sky as the Earth turns on its axis. The centre of all of the circles marks the position of the north celestial pole. Mauna Kea is only 20° north of the Earth's equator, so the pole appears very low down in the sky.

Polaris is always visible in the northern hemisphere, and so it has been used through the ages for navigation by sailors. You can find it by following the line which joins the two stars in Ursa Major which are furthest from the long tail. As a double check, Polaris

Orion, the hunter. At the centre are three stars forming 'Orion's Belt.'

The Stars

Heavenly Bodies

The Stars

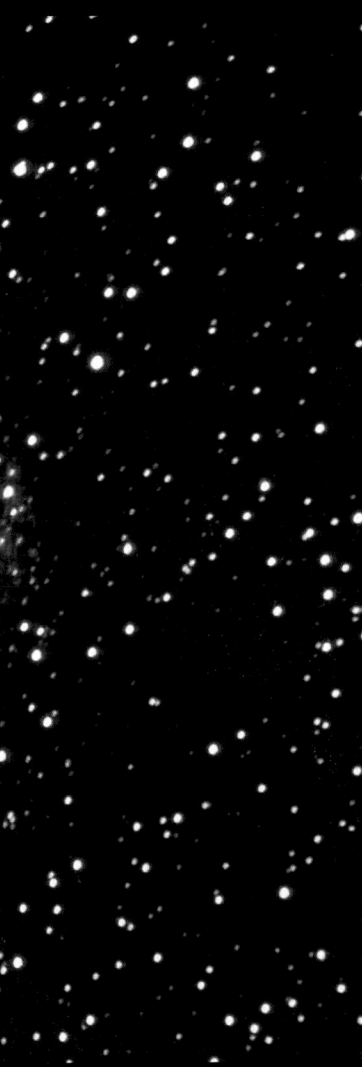

makes the tip of the tail in the constellation Ursa Minor.

For most southern observers, Orion will still be visible at the start of the year, but the Southern Cross, high in the sky, will usually be your best starting point. Nearby are the bright star Alpha Centauri, one of the Sun's nearest stellar neighbours, and close by in the sky but 50,000 times further away in space, the stunning Large Magellanic Cloud, a companion galaxy to our own Milky Way.

The Large Magellanic Cloud is one of two irregular, dwarf galaxies which are orbiting companions of our own galaxy, the Milky Way. The LMC is some 150,000 light years distant. Together with the Small Magellanic Cloud, it is named after the Portuguese navigator Ferdinand Magellan, who first described them during his circumnavigation of the globe in 1519.

Heavenly Bodies

The Ecliptic

The brightest objects in the night sky after the moon are the planets Venus, Mars and Jupiter. They are much closer than the stars and so move against the background constellations. For this reason their positions can not be marked on a conventional star chart. However, the planets do all orbit the Sun in roughly the same plane, and so when viewed from the Earth they all appear to move along the same track as the Sun. This path, which encircles the entire sky, is called the Ecliptic and it is marked on star charts.

In order to find out where a planet is on the Ecliptic you need to know where it is on its orbit on the date that you are out looking for it, and for that you need to have a map like the one on page 28. It shows where some of the brighter planets will be on 1 January 2000.

The lines plot the planets' orbits, as they would appear when viewed from far above

The Ecliptic

The planets of our Solar System are (bottom left to top right): Pluto, Neptune, Uranus, Saturn, Jupiter, Mars, Earth, Venus and Mercury.

Heavenly Bodies

the its daily rotation in an anti-clockwise direction. The numbered circles show where the planets will be at three-month intervals through the year 2000; they are equally spaced so you can add to them for the following years. The three-monthly positions for Venus do not appear in numerical sequence because Venus is much closer to the Sun, and moves one and a half times around its orbit in a year. The planets are not drawn to scale because they would then be far too small to see – for example, the distance from the Earth to the Sun is some 150,000,000 km, but the Earth is only about 13,000 km in diameter.

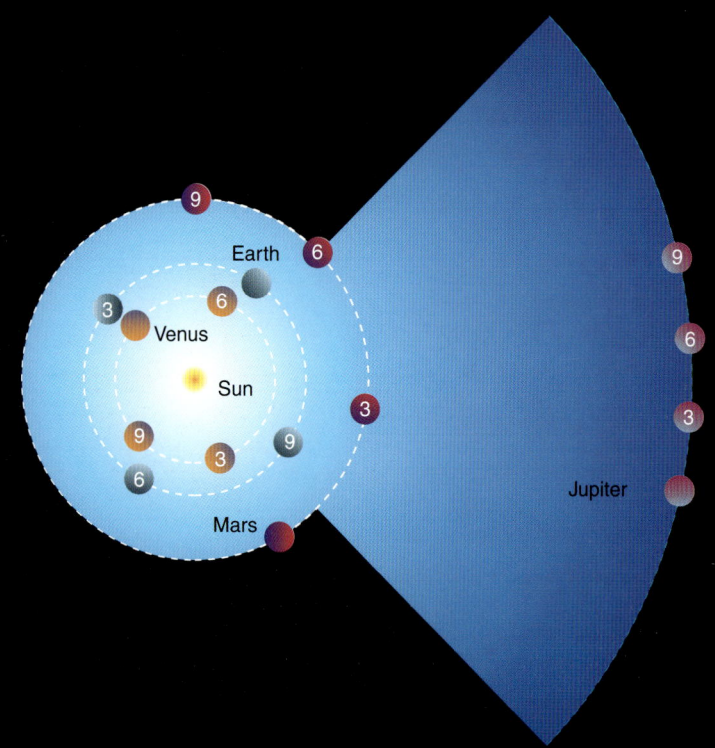

Venus

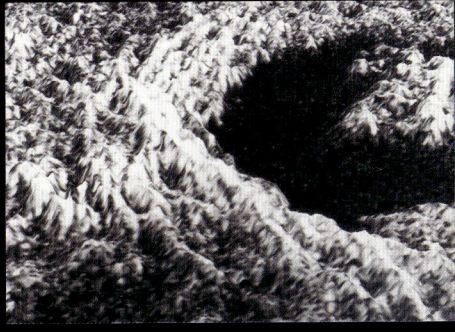

The Golubkina crater on Venus, which is about 34 km in diameter.

The planet Venus is the brightest object in the night sky after the Moon. It can always be found in the same part of the sky as the Sun when viewed from Earth, and this comes about because it orbits more closely to the Sun than the Earth. As a result it can best be seen early or late in the day – which has given rise to its traditional name of the Morning Star or the Evening Star, as the brightness of the Sun makes it difficult to see at any other time of day. In the summer of 1999 Venus can be seen around sunset, and in the winter of 2000 it can be seen at dawn. In the intervening period its orbit takes it on the far side of the Sun to the Earth, and it cannot be seen with the naked eye at all. You should never look directly at the Sun or anything too close to it, because the intense light could do irreparable damage to your eyesight.

Now that we learned how to find objects in the night sky, let's see what astronomers have found out about them, starting with our own Solar System.

Photo using a blue filter to enhance the clouds that swirl around Venus.

THE ORIGINS OF THE SOLAR SYSTEM

- **Nebulae**

- **Comets, Meteors and Asteroids**

- **The Discovery of New Planets**

CHAPTER 2

The Origins of the Solar System

Nebulae

One of the biggest questions facing astronomy is that of how the Solar System came into being. A full picture has started to emerge only very recently thanks to two things, the dawn of the age of space exploration, and the construction of a new generation of large ground-based telescopes. Observations from sites like the Mauna Kea observatory in Hawaii have given us a view of the formation of planetary systems like our own with unprecedented detail.

Planetary systems are believed to form from the giant clouds of dust and gas that are seen throughout the Universe in galaxies like our own. The nearest, and therefore the best studied region of star formation is in Orion; you can just see it with the naked eye as the smudge below Orion's belt. This nebula, which is about 500 parsecs (see p.41) from the Earth, is where newly formed bright stars are shining on their parent cloud, heating it up until it glows. The picture below shows the view that a big telescope gives us of the Orion Nebula. Two million years or so ago the view would have been quite different. The bright

The Orion Nebula, a bright cloud of gas and dust where stars are being born.

Telescope domes on the 4200m summit of the dormant Mauna Kea volcano in Hawaii, one of the world's best sites for astronomical observations.

nebula would not have existed, and instead the picture would have shown the implacable black wall of a giant molecular cloud. Molecular clouds can stretch for tens of parsecs in all directions; they contain the mass of thousands of stars in the form of gas and dust, balanced on a knife edge of stability between the compressive force of gravity and the expanding force of the gas pressure. Invisible to the eye, but important to the formation of stars, weak lines of magnetic force thread their way through the clouds, providing a skeleton to help support them against the force of gravity.

A million years ago, a parcel of gas and dust deep inside the Orion molecular cloud that was slightly more dense and cool than its surroundings became unstable. The force of gravity overcame the pressure of the cold gases and the parcel began collapsing. As it got smaller, it began rotating faster and faster, spinning a disk of

The red areas in this molecular cloud show where stellar formation is taking place.

material out from its equator and twisting the original cloud's threads of magnetic force into a tight rope. Within a few tens of thousands of years, the central core had become dense enough and hot enough for nuclear fusion to occur, and a star was born.

The explosion into life of a star has a dramatic effect on its surroundings. New stars burn their way out of their parent cloud and leave a huge cavity. Young stars are seen to fling gas many parsecs out into space along the twisted magnetic ropes that extend from their equatorial disks.

These violent events eventually run out of steam, leaving a star like our Sun settling down to a life that may last for several billion years.

The birth of planets

The next stage, the formation of a system of planets from the primeval disk of gas and dust, has been studied by theoreticians for years. Until the recent advent of very powerful telescopes (see chapter 4) they have only had our own Solar System as the test-bed for their models. This has restricted their observations to the variations found in just a

Nebulae

single, small solar system. The planets in the Solar System come in two types. The first type is the terrestrial planets, Venus, Mars, Mercury and our own Earth and its Moon, which orbit quite close to the Sun, and are mainly composed of rocky material. Further out from the Sun are the gas giants, Jupiter, Saturn, Neptune and Uranus, which are much bigger and heavier than the terrestrial planets, but are mainly composed of the light elements hydrogen and helium, small quantities of which are fond on the Earth as gases.

The current picture of Solar System formation has the rocky, solid particles condensing near to the parent star where they can survive the intense heat, while icy grains form further out where it is colder. Over millions of years, the particles coalesce into larger and larger bodies, eventually with one body coming to dominate in any given orbit, sweeping up the available debris like a snowball growing as it rolls down a hillside. The high mass of the gas giants comes about because they have much bigger orbits to sweep clear of debris. There is also a lot more of the volatile elements hydrogen and helium available in the circumstellar disk than there is of the heavier elements like iron and silicon that make up the terrestrial planets.

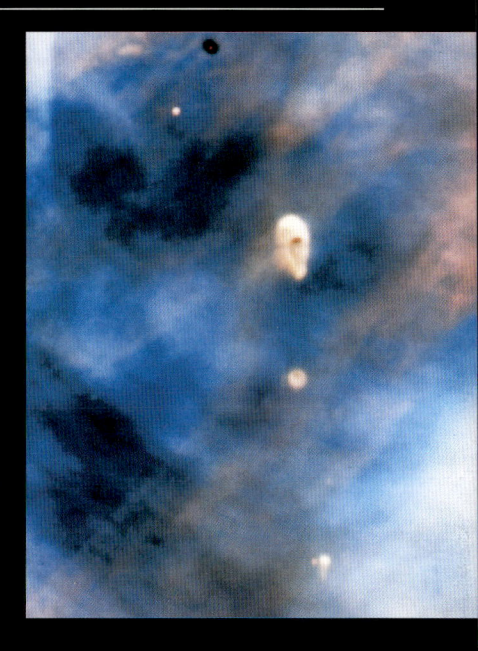

Protoplanetary disks, or proplyds, which surround recently born stars. They might evolve and, following a process of condensation, form planetary systems like our own Solar System.

The Origins of the Solar System

Comets, Meteors and Asteroids

In the next few hundred million years following their birth, the planets would have suffered a concentrated bombardment from space, as the comets and asteroids left over from the original cloud of planet forming material crashed into them. Now, over four billion years after the Solar System was formed, we are still visited by the occasional comet, an icy lump that has survived all this time, way out from the Sun on the edge of interstellar space, only to be toppled from its orbit by a passing star. Plummeting into the warmth of the inner Solar System, the tails of comets are dust and gas being boiled away from the frozen nucleus, streaming away from the Sun like long hair blowing in the Solar wind. If they become captured by a planet's gravitational pull,

The impact of the 1 km diameter comet, Shoemaker-Levy, on Jupiter.

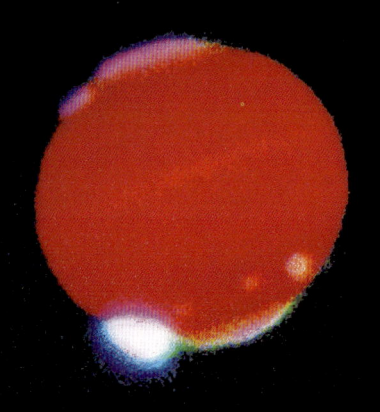

Comets. Meteors and Asteroids

Halley's Comet, in the constellation of Sagittarius. The comet is a lump of ice and dust that completes a highly elliptical orbit around the Sun every 76 years. The comet's nucleus (the brightest part) is about 16 km long, while the tail blazes behind it for millions of kilometres.

comets are doomed eventually to break up into fragments which follow the same orbit. When the Earth intercepts the orbit of a 'captured' comet, we see these fragments streaking across the night sky as meteor showers. The impact of the comet Shoemaker-Levy on Jupiter in 1994 was a spectacular reminder of how lucky we are to live on a small planet deep in the protective gravitational well of the Sun, where we are relatively safe from the impact of comets and asteroids.

The Discovery of New Planets

In November 1995, the astronomers Michel Mayor and Didier Queloz of the Geneva Observatory in Italy, discovered tiny variations in the velocity of the star 51 Pegasus by making precise measurements of its spectrum. Something was making it move towards, and then about two days later away from, the Earth. The only possible explanation was that it was being tugged by the gravitational pull of a planet, roughly as massive as Jupiter and taking only 4.2 days to orbit the star. This was the first verifiable discovery of a planet outside our solar system, and since then over a dozen others have been found. These new planets can only be detected because they are very big (in other words they are gas giants) and very close to their parent stars, two things which should not happen together according to our simple picture of planet formation.

An artist's impression of the planet orbiting the star, 51 Pegasus. Two moons can be seen against the planet. The star is a Sun-like star some 42 light years from Earth. The planet is about half as massive as Jupiter, and orbits only 7 million km from the star, the equivalent of only a sixth of Mercury's orbit around the Sun. The surface temperature is probably over 1,000°C, and the rocks of the planet at least partly molten.

The Discovery of New Planets

Why this should happen is unclear; maybe the planets were formed further away from their parent stars and moved into their current orbits later on. Whatever the reason, this exception to the rule shows the way in which science never gives us a complete

The Origins of the Solar System

Artist's view of a hypothetical planet with seas and a sulphur-based atmosphere.

and finished picture. Each discovery just makes things a little clearer, and if you are lucky it points to new avenues of research.

The hunt is now on for more direct evidence than these star-wobbles of other extra-Solar planets. A camera called SCUBA which uses microwave light is helping in this quest. Microwave light has a wavelength hundreds of times longer than visible light.

The Discovery of New Planets

The wavelength of the light is important because it is an indicator of temperature. The cooler something is, the longer the wavelength of the light it emits. Our eyes are adapted to detect light from the Sun, which is at a temperature of about 6,000°C, but SCUBA's electronic eyes are best for seeing things at an icy –200°C. The pictures taken by SCUBA may be able to show us what our own Solar System looked like soon after it formed.

Fomalhaut, a star which is only two hundred million years old and much younger than our own Sun, has recently been discovered by SCUBA to have a ring of dust around it. This is what you would expect to find in a newly-forming planetary system. Fomalhaut is a bright star in the southern sky which you can find on your star globe. In the northern hemisphere, the bright star Vega in the constellation of Lyra has had the same sort of dusty ring discovered in orbit around it.

The world's great telescopes are helping astronomers to find new planetary systems every few months, and the existence of planets around stars is now believed to be quite commonplace. The distance from the Earth to the stars is so huge that it is unlikely that we will be able to visit them ourselves, but we have been able to send robots to investigate the planets in our own Solar System, and, as we will now see, they have proved to be fascinating places.

ASTRONOMICAL DISTANCES

You have to get used to mind-boggling numbers if you take up astronomy. Things can be much much bigger or hotter or faster than we are used to in our everyday lives. On the Earth we measure distances in kilometres or sometimes in miles, but in astronomy the kilometre is too small, and distance is more usefully measured in parsecs. One parsec is equal to an inconveniently large 30,000,000,000,000 km. It can be broken down into a more manageable number, with one parsec also equal to 200,000 times the distance from the Earth to the Sun. The parsec really comes into its own beyond the Solar System, where it is comparable to the typical distance between stars in our part of the universe.

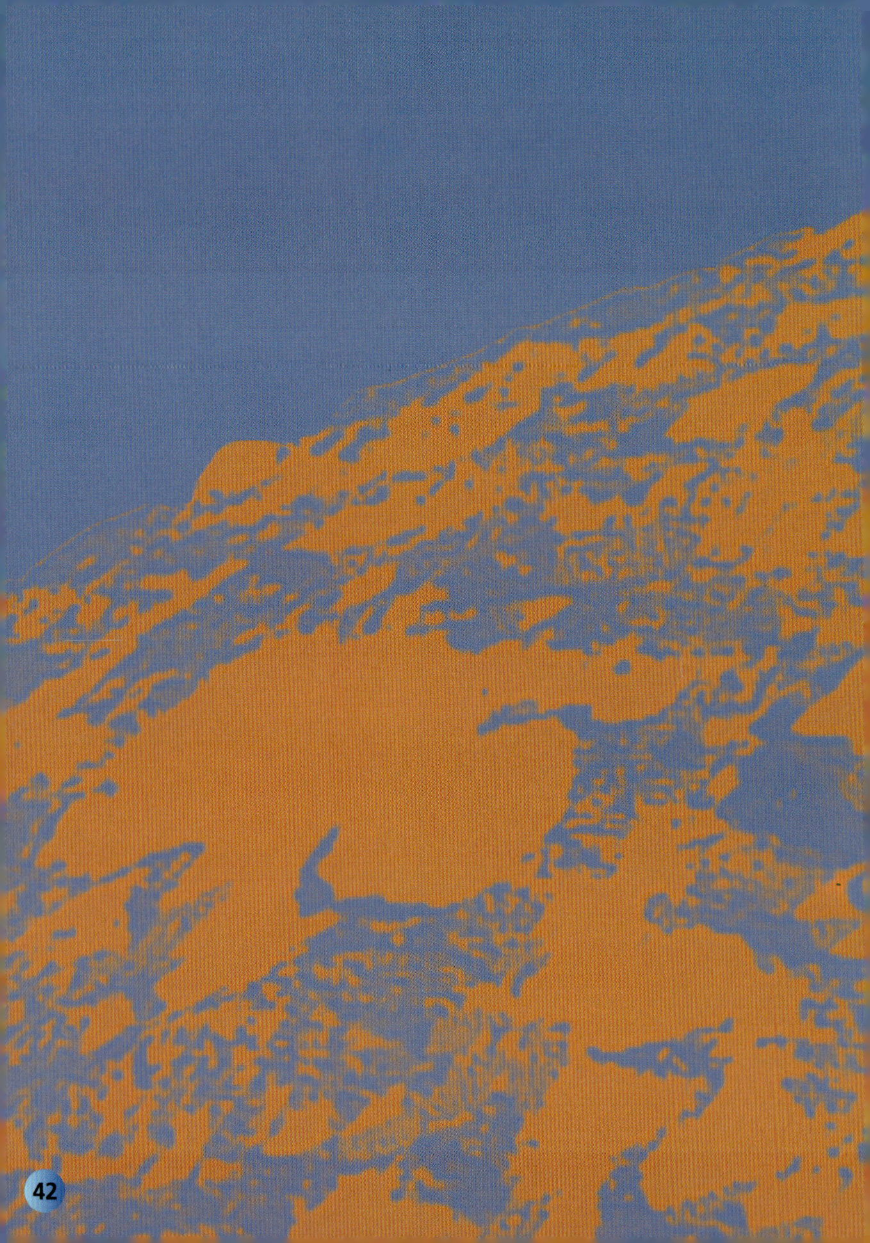

EXPLORING THE SOLAR SYSTEM

- **Exploring in the 21st Century**
- **Mars**
- **Jupiter**
- **Saturn**

CHAPTER 3

Exploring in the 21st Century

However big your telescope is, it can never tell you as much about a far-off planet as you will find out by visiting it. Thanks to the United States' National Aeronautical and Space Agency (NASA), a few dozen spacecraft have been sent to explore the furthest reaches of the Solar System over the past thirty years, and they have given us a totally new understanding of what the planets are really like. Both NASA and ESA (the European Space Agency) have exciting plans for more ambitious missions. In the next few years there are plans for a spacecraft to intercept the orbit of a comet and take samples, a series of landers to study the surface of Mars, and a probe to land on Titan, a moon of Saturn.

It is a safe bet that future space exploration will keep turning up surprises about our planetary neighbours. Titan is believed to have much in common with the early Earth and so it has been made the target of a mission by ESA and NASA. In 2004, ESA's Huygens probe is planned to separate from the Cassini spacecraft over Titan, and parachute down through the thick clouds of methane that hide the surface from view. If it survives the trip to reach the surface, you can be sure that Huygens will find something as surprising as was the ring of dust around Vega.

Space exploration began in 1959 when Russia's Luna 2 probe took three days to make the voyage to the Moon where, entirely according to plan, it crashed. This pioneering success helped to spur the United States into a race to the Moon that culminated in the manned Apollo landings between 1969 and 1972. The 300 hours spent by 12 astronauts on the lunar surface are still the sum total of mankind's direct exploration of another world. In the last thirty years manned exploration has been directed towards space-stations in close orbit about the

Artwork of the Cassini spacecraft approaching Titan.

Earth, and a manned mission to Mars is still just a distant dream.

Exploring the Solar System

Mars

Mars has been a target for unmanned exploration ever since Mariner 4 made its flyby on 14 July 1965. It is the planet most like the Earth that we know, and there is still a small chance that it harbours some primitive form of life.

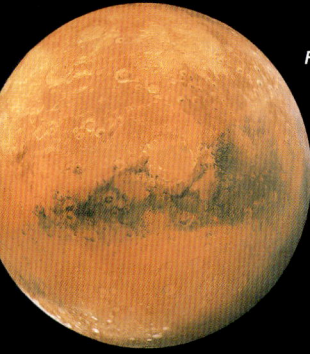

Full view of Mars centred on the large Schiaparelli impact crater.

Mars is smaller than the Earth; the atmosphere at the bottom of its deepest canyons is a hundred times less thick than at the surface of the Earth, and it only contains a tiny trace (about 0.15%) of oxygen. However, the pictures taken from space by the Mariner and Viking missions show a landscape carved by water erosion, providing startling and completely unexpected evidence

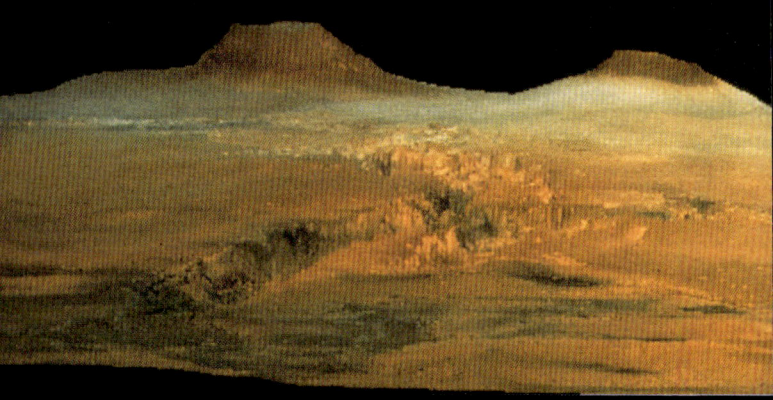

Mars

that liquid water has flowed on the Martian surface in the distant past. This water is now frozen in icy reservoirs beneath the planet's surface and in the polar ice caps. The planet's surface only reaches the freezing point of water at the height of Martian summer; usually temperatures are a chilly −50 to −100°C, with Mars' polar ice caps containing frozen carbon dioxide as well as water.

Mars will be a spectacular place to visit when people do finally manage to make the journey. On the Earth, the crustal plates that we live on float on an interior of molten rock, and over the geological ages these slide into, under and over each other, making our highest mountain ranges endure for only a few tens of millions of years. On Mars, the crust locked solid when the planet was young, over three billion years ago. It was unable to relieve the pressure of upwelling molten magma by moving aside, and instead buckled and cracked to make extraordinary features like the Valles Marineris, a 4000-km long chain of 5-km deep canyons.

Mars also has volcanoes. The highest volcano on the Earth is Mauna Kea, which is on the latest of the Hawaiian islands to have grown above a hot spot in the crustal plate which is moving northwards across the Pacific. On Mars, the hot spots do not move, and so volcanoes just grow bigger and bigger. Olympus Mons is the biggest volcano in the Solar System; at 24 km it is twice as high as Mauna Kea, and it has a 6-km high cliff encircling its base.

The giant Valles Marineris canyon system (left foreground), and the three Tharsis volcanoes in the distance.

47

Jupiter

The most spectacular active volcanoes are found much further out from the Sun than Mars, beyond the asteroid belt (more debris left over from the formation of the Solar System), on a moon of the gas giant planet Jupiter, called Io. Active volcanoes were the last thing the mission scientists on Voyager 1 were expecting to see as the spacecraft approached Jupiter in 1979, because Io should be too small to have a molten interior. They were prepared for the impact craters which are seen on all of the terrestrial planets, but instead its surface looked yellow and blotchy, covered in sulphurous compounds. After a few days a spectacular volcanic eruption was seen, spewing sulphur hundreds of kilometres up into space.

Io's molten interior is now believed to be stirred into its volcanic fury by the tidal force of Jupiter. Jupiter is about the same

Photos of Jupiter taken four months apart by the Voyager 1 and 2 spacecrafts, showing how the planet's atmosphere is constantly changing.

Jupiter

Mosaic of Voyager spacecraft photographs of Jupiter and its four Galilean moons: Callisto (bottom right), Ganymede (bottom left), Europa (centre) and Io (top left).

The small satellite orbiting Jupiter is Io, the innermost of the four Galilean moons of Jupiter.

distance from Io as Earth is from our Moon, but Jupiter is over 25,000 times as massive as the Moon, so the tides it causes inside Io are 25,000 times stronger than our own. Jupiter and its moons are a bit like a miniature Solar System. All four moons were discovered by Gallilieo n 1610.

Io is analogous to the terrestrial planets, with a higher proportion of heavy, rocky elements in its makeup, while further out from Jupiter the moons Europa, Ganymede and Callisto are progressively less dense, with thick mantles of water ice encasing a rocky core. There is a possibility that Europa might have a layer of liquid water many kilometres down beneath its cracked icy surface, warmed by

Jupiter

the same tidal heating that powers Io's volcanoes. If so, it could be another potential site for life outside the Earth, but it will be a long time in the future before we will be able to find out.

Jupiter itself has a beautiful appearance thanks to the complex chemical compounds which colour the swirling bands and spots that flow across its visible disk. Jupiter does not have a solid surface. If you were to dive beneath its top layer of ammonia clouds, the atmosphere would get warmer and thicker until eventually at great depths it would smoothly take on the properties of a liquid.

With no mountain ranges or hot sunshine to break up the smooth weather patterns, Jupiter's atmosphere is like a gigantic wind tunnel, able to support features like the Great Red Spot, a vortex twice the size of the Earth which has survived for hundreds of years.

The other gas giants, Saturn, Uranus and Neptune are similar to Jupiter, but their cloud patterns are less distinct and barely visible from the Earth. The bands and vortices seen on Jupiter, however, were still visible to the two Voyager spacecraft when they passed by on their journey through the Solar System.

Jupiter's Great Red Spot (top right corner) and the cloud formations to the south and west of it. The Great Red Spot is believed to be a storm system, several times the size of the Earth, and over 200 years old.

Exploring the Solar System

Saturn

Saturn is famous for its spectacular ring system, which, incredibly, is less than 100 m thick but can still be seen through a small telescope from the Earth. All of the giant planets have rings. They are not solid, but made up of millions of tiny particles of water ice, with gaps and braids woven into them by the gravitational disturbances of their planets' moons. The origin of the rings is still unclear, but they appear to be the transitory remains of captured icy comets or satellites, torn apart by the planets' tides.

Over its lifetime of billions of years, a planet like Saturn may be decorated by a ring on several occasions, and each one can be seen as a beautiful example of how the Solar System continues to change.

Colour enhanced photos of Saturn and its famous rings, taken by the Voyager 2 spacecraft in July 1981, when it was 43 million km from Saturn. The rings are composed of particles of ice.

Saturn

54

BEYOND THE SOLAR SYSTEM

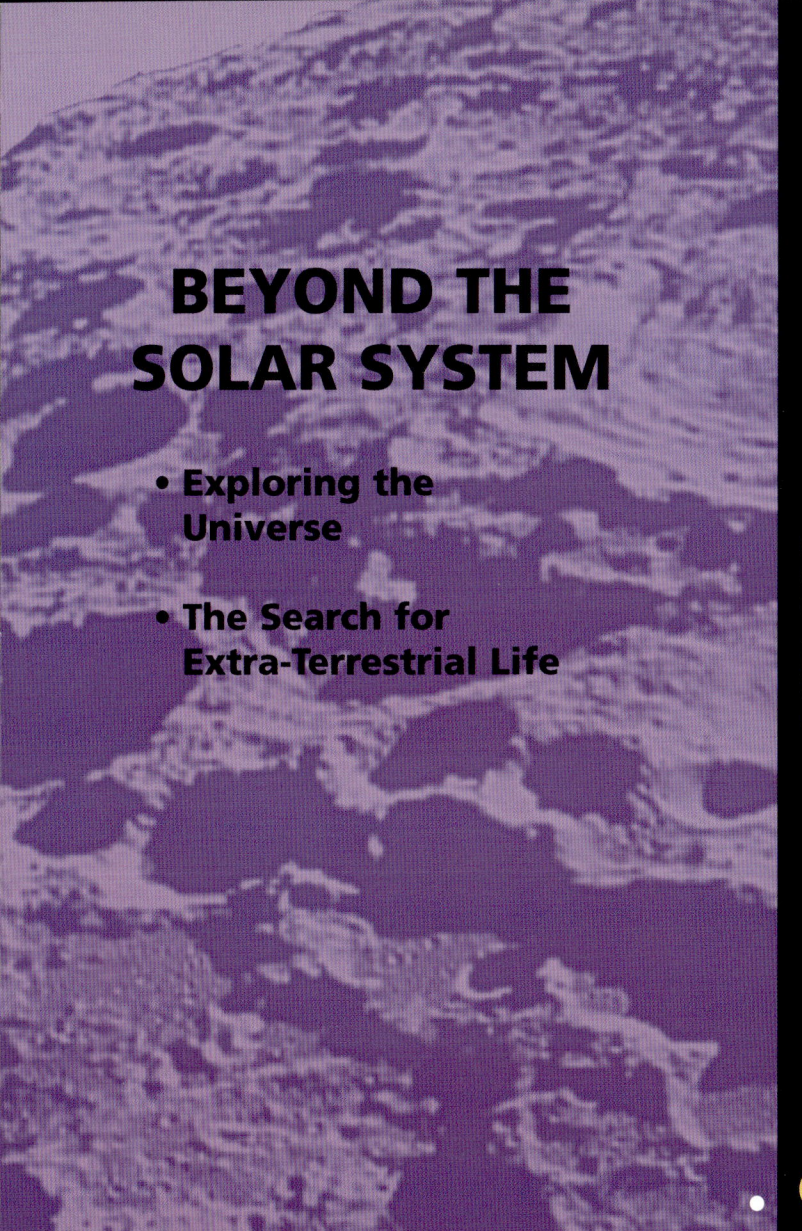

- **Exploring the Universe**

- **The Search for Extra-Terrestrial Life**

CHAPTER 4

Beyond the Solar System

Exploring the Universe

When we discussed in Chapter 2 how the Solar System had its origin four and a half billion years ago in a cloud of gas and dust, we avoided the obvious question of where the molecular cloud came from.

The answer is that interstellar clouds are formed from the smoke of old stars. But the questions just pile up. Where did the first stars come from? Why do stars cluster together in their millions to form galaxies? Finally behind them all is the big one. Has the Universe of stars and galaxies existed forever, and will it carry on into the future forever too? The search for answers to these questions is the subject of a branch of astronomy called cosmology.

The Universe is very big. Even parsecs are too small to measure cosmological distances unless we use them a million at a time. The shape of the Universe is also a factor to consider. The fastest thing in the universe is light. It can get from the Moon to your eye in just over a second, and from the Sun to your eye in eight minutes. The light entering your eye when you look at the Orion nebula has been travelling for 1500 years. As a result, distances are often calculated in light years (the distance light travels in one year).

The further away a galaxy is, the further back in time you are seeing it, and the light from some of the galaxies in the Hubble

Edwin Hubble at the Mount Wilson Observatory, USA.

The vast Hubble Space Telescope seen shortly before deployment from Shuttle Endeavour, December 1993.

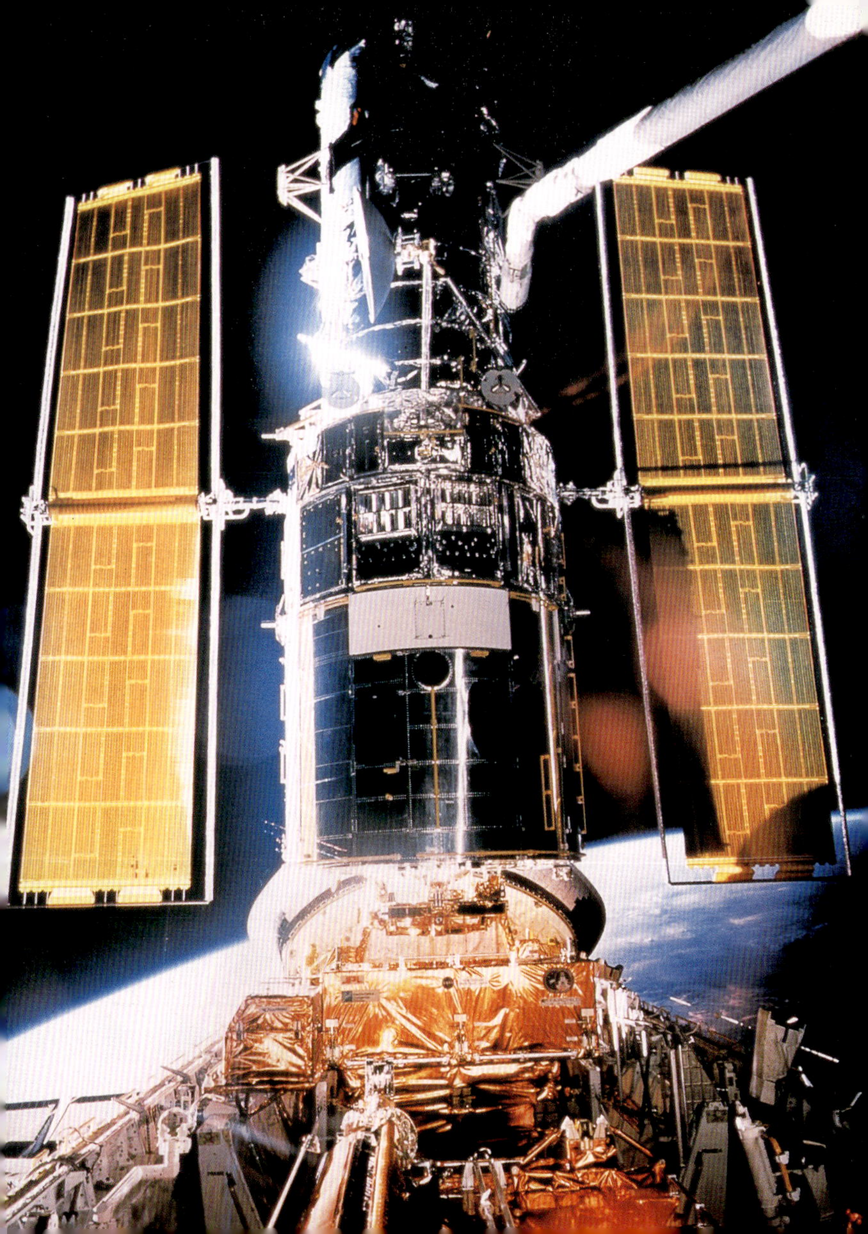

An astronaut servicing the Hubble Space Telescope in December 1993.

Deep Field started its journey towards us before the birth of the Sun. It is called the Hubble Deep Field because it was discovered by the Hubble Space Telescope, but its name also honours Edwin Hubble, an astronomer who made an astonishing discovery about distant galaxies. He found that the colour of the light from these galaxies has shifted towards the red end of the spectrum, and the more distant they are, the bigger the shift. This phenomenon, called the cosmological red-shift, is caused because the galaxies are moving apart; the Universe is expanding.

The implication of the galaxies moving apart from each other is that long ago in the past they were all a lot closer together. In fact, most cosmologists now agree that about 15 billion years ago all of the matter in the Universe was condensed into one incredibly hot place. Since the explosion of this matter, the Big Bang, it has been flying apart, expanding and cooling to form the stars in their galaxies that we now find around us.

The most distant object ever seen has a red-shift which corresponds to a recession velocity very close to the speed of light. It is not a galaxy, but is the wall of glowing gas that existed just after the Big Bang. Called the cosmic background, its red-shift is so big that its spectrum has been shifted from the visible (back then it would have been brighter

Exploring the Universe

than the surface of the Sun), through the red and infra-red, deep into the microwave part of the spectrum. This is where the idea of the shape of the Universe comes in, because the cosmic background can now be seen in all directions around the sky, even though it was created when the whole Universe was only about 300,000 years old, and so just a tiny fraction of its current size.

One of the hardest cosmological problems to explain is the fact that when the Universe was formed it must have been very uniform and smooth, but today it is lumpy, with stars clustered into galaxies which form clusters of their own. On a massive scale the Universe has the appearance of a foam, with voids of relatively empty space between sheets and threads of matter.

At the centre of the debate about the destiny of the Universe is the question of how much the Universe weighs. If it is heavy enough, then gravitational attraction will eventually slow down and stop the galaxies which are currently flying apart, and maybe even bring them back together in the distant future in a great cosmic crunch. However, it is not easy to weigh the Universe. When astronomers have tried to understand how galaxies move in galaxy clusters, they have found that their results could only be explained by the presence of large amounts of invisible, 'dark matter'.

The current view has it that there is not enough dark matter in the Universe to keep it from expanding forever. Far into the future, our galaxy will be left on its own, with all of its stars left as burned-out cinders. Thankfully, we have a few billion years to enjoy before the end comes.

Hundreds of distant galaxies in the Hubble Deep Field.

Astronomy

The Search for Extra-terrestrial Life

As far as we know, the only place in the universe where life exists is on our own planet, the Earth. We have already seen that the other planets and moons in our Solar System are more or less hostile to life as we know it. With the possible exception of Mars, their surfaces are either too hot or too cold, and they are all far too dry. If there is a trace of life on one of them, it will almost certainly be both very primitive and very exotic.

It is the liquid water in the Earth's rivers and oceans, and falling from the sky as rain, that has allowed life to flourish and spread into every corner of our planet, to the extent that the living green forests and grasslands of our major continents can even be seen in pictures taken from space, making the Earth a unique jewel among the nine planets that orbit our Sun.

We would not expect to find anything we would recognize as life on any of the extra-solar planets that have been discovered so far. Like the one orbiting 51 Pegasus, they are all too massive or too close to their parent star to hang on to liquid water, and so they will not be remotely like the Earth. However, they do give us the encouraging indication that many stars have planets, and so we can have some expectation

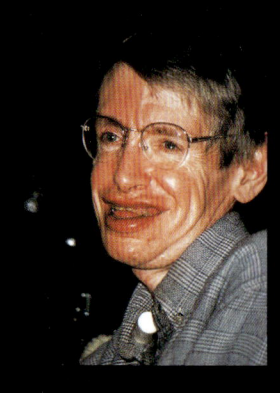

Professor Stephen Hawking whose books on black holes and the 'Big Bang' have popularized astronomy.

The Search for Extra-Terrestrial Life

that there will be more habitable terrestrial planets orbiting nearby stars too. The question is whether we can find them.

The wobble given by terrestrial planets to their parent stars will be tens of thousand of times smaller than that given by gas-giants, and so the techniques used by Mayor and Queloz (see page 38) will have little chance of success. There are other ways to find extra-solar planets, however. One of them is to measure carefully the amount of light coming from nearby stars over a long period of time, say a few years. If one of the stars has an Earth-like planet orbiting it, and if that planet's orbit happens to take it between us and the star, it will cut off (occult) a tiny fraction of the star's light, about 1 part in 15,000. It should be possible to measure this tiny dip in the starlight using a space-borne telescope, and if enough stars are measured, some of them are bound to have planets with just the right sort of occulting orbit. However, it still won't tell us whether the planets have life.

The Earth's rivers, oceans and rainfall have enabled life to spread over the whole planet.

The detection of signs of life on Earth-like planets orbiting nearby stars is the goal of ESA's Darwin Mission. NASA has a similar mission, called the Terrestrial Planet Finder. The plan is to launch a flotilla of up to five modestly sized telescopes sometime in the next twenty years or so, and fly them to a dark place in our Solar System. When they arrive there, they will then be positioned very precisely in a formation that will allow them to mimic a single telescope with a mirror 100 m or more in diameter, at least with respect to the amount of detail they can see in a distant star system.

The telescopes will then search nearby stars for light emitted at thermal infra-red wavelengths, looking for warm, terrestrial planets. Darwin should be able to find a planet like our own Earth at a distance up to 20 parsecs if one exists, and there are hundreds of stars within that range.

Darwin won't be able to show us the seas and forests on any of the planets it discovers; in fact the entire planet will only appear as a fuzzy dot. To find out if there is life present, the fuzzy dot will have to be analyzed spectroscopically, in other words, Darwin must measure precisely what colour it is. Objects have characteristic colours because they absorb or emit certain wavelengths of light more than others. The green of leaves and the colour of your eyes come about because they each have a different chemical makeup. The colour that is believed to mark out a planet as being inhabited by life is the bright thermal infra-red colour of ozone gas.

Ozone is a molecule which is made high up in the Earth's atmosphere from the oxygen which we must breathe to live. Oxygen, in turn, is made by the plants, trees and algae that cover the Earth's surface. The idea behind Darwin is simple. If the fuzzy dot that marks a

The Eureca satellite deployed from Shuttle Atlantis. Eureca carries various experiments in life and materials science and radiobiology.

This photo taken by the Hubble Space Telescope is the first direct evidence of an extrasolar planet.

distant planet glows with the colour of ozone, then it must have an atmosphere full of oxygen, and there must then be some life-form on the planet that is making that oxygen. It may look stranger than the strangest creature you could imagine, but it will share our basic chemistry.

Of course, there may be life out in space that won't draw attention to itself by filling its planet's atmosphere with oxygen and so it would never be detected by Darwin. In our search for extra-terrestrial life, it makes sense to start by looking for something familiar; we can always carry on the search for more bizarre life-forms later on.

During its mission, Darwin will be able to search over a hundred nearby star systems for signs of Earth-like life. If it finds nothing the search will not be finished, but we will have learned that the Earth is truly special, and should be valued as such. If Darwin does find life on just one planet, the likelihood is that we must have distant relatives on millions of planets across the galaxy.

In a few years, when you go outside to watch the stars, there is a good chance that you will be able to pick out a point of light and know that you are looking at a planet like the Earth, where strange beings live. Perhaps they too will be advanced enough to be enthralled by the stars in the night sky. They might even be looking back at you.

Astronomy

Index

Numbers in italics refer to pictures.

Aldrin, Buzz 18
Alpha Centauri, 25
Apollo 11, *17*, 44
Armstrong, Neil 16, *16*, 18
Asteroids, 36, 37, 48
Astronauts, 44–5

Big Bang, 58, *60*

Callisto, 50
Cassini, 44, *45*
Cassiopeia, 21
Colour, spectroscopy 62
Comets, 36–7, 44, 52
Constellations, 21–25. *See also* Cassiopeia, Lyra, Orion, Plough, Southern Cross, Ursa Major, Ursa Minor.
Copernicus, Nicolaus, 11, *11*
Cosmic background, 58–9
Cosmology, 56–59

Dark matter, 59
Darwin Mission, 61–3
Distances, 41, 56–8

Earth, *18*, 26, *27*, 35, 47, 60, 61, *61*, 62
Eclipses, *14*, 14–15, *15*,
Ecliptic, 26–8
Energy, solar 12–14
ESA (European Space Agency), 44, 61
Europa, 50–1
Extra-terrestrial life, 60–3

Fomalhaut, 41

Galaxies, 25, 32, 56–8, 59
Galileo, 50
Ganymede, 50
Geneva Observatory, 38
Gravity, 11–12, 16, 33, 36, 37, 38,59
Great Red Spot, 51, *51*

Halley's Comet, *37*

Hawking, Stephen *60*
Helium, 13, 35
Hubble, Edwin 56, 58
Hubble Deep Field, 56–8, 59
Hubble Space Telescope, 56–57, 58, *58*, *63*
Huygen's probe, 44
Hydrogen, 13, 35

Io, 48–50, *49*, *50*, 51
Iron, 35

Jupiter, *18*, 26, *27*, 35, *36*, 37, 48–51

Large Magellanic Cloud, *24–25*, 25
Life, extra-terrestrial 60–3
Light, 40–1, 56–8, 61, 62
Luna 2 probe, 44
Lyra, 41

Magellan, Ferdinand 25
Magnetism, 34,
Mariner 4, 46
Mars, 26, *27*, 35, 44, *46–7*, 60
Mauna Kea Observatory, *33*
Mayor, Michel 38
Mercury, *27*, 35
Meteor showers, 6, 37
Microwave light, 40–1
Milky Way, 4, 5, *4–5*, 25
Molecular clouds, 33–4, *34*, 56
Moon, 10, 12, 14, 16–19, *18*, *19*, 35, 44, 56

NASA (National Aeronautical and Space Agency), 44, 61–2
Nebulae, 32–5
Neptune, *27*, 35, 51
Newton, Isaac, 11
Nuclear fusion, 12, 34

Olympus Mons, 47
Orion, 21, *23*, 25
Orion nebula, *32*, 32–4, 56
Oxygen, 62–3
Ozone, 62–3

Parsecs, 32, 33, 34, 41, 62

Pegasus 51, 38, *38–39*, 60
Planets, 26–9, *27*, 32, 34–5, 36, 38–41, *40*, 46–53, 60–3, *63*
Plough, 21, *21*
Pluto, *27*
Polaris, 7, *22*, 22–5,
Proplyds, 35
Protoplanetary disks, *35*

Queloz, Didier 38

Red shift, 58
Rings, Saturn 52, 52–3, *53*

Sagittarius, *37*
Satellite, *62*
Saturn, *27*, 35, 44, 51, 52–3
SCUBA, 40–1
Shoemaker-Levy, 36, *36*, 37
Shuttles, *57*, *62*
Silicon, 35
Solar System, 32, 34–5, 36, 41, 44, 56, 60, 61–2
Solar wind, 14, 36
Southern Cross, 25
Space stations, 45
Stars, 5, 20–5, 32, 33, 34, 38, 39, 56, 58, 59, 60–1, 62, 63
Stonehenge, *10*, 10–11
Sun, 5, *6*, 7, 10–15, *11*, *12*, *13*, 29, 35, 36,41, 56

Telescopes, 6, 32, 33, 41, 57-58, *57*, *58*, 61–2
Terrestrial Planet Finder, 61
Titan, 44, *45*

Universe, 56–9, 60
Uranus, *27*, 35, 51
Ursa Major, 21, *21*, 22
Ursa Minor, 25

Valles Marineris, *47*
Vega, 41, 44
Venus, 26, *27*, 28, *28*, 29, *29*, 35
Viking, 46
Volcanoes, 47, 48
Voyager 1 and 2, 48, *48*, *49*, 51, *52*, 53